MW01356411

How to Use This Book

Get it dirty.

Smudge the pages.

Write in the margins.

Sprinkle it.

Wrinkle it.

And take it into your garden.

This is a handbook, one we hope you'll tuck into your gardening bucket among your trowels and spades.

Read it.

Refer to it.

Keep notes.

And make it your very own book.

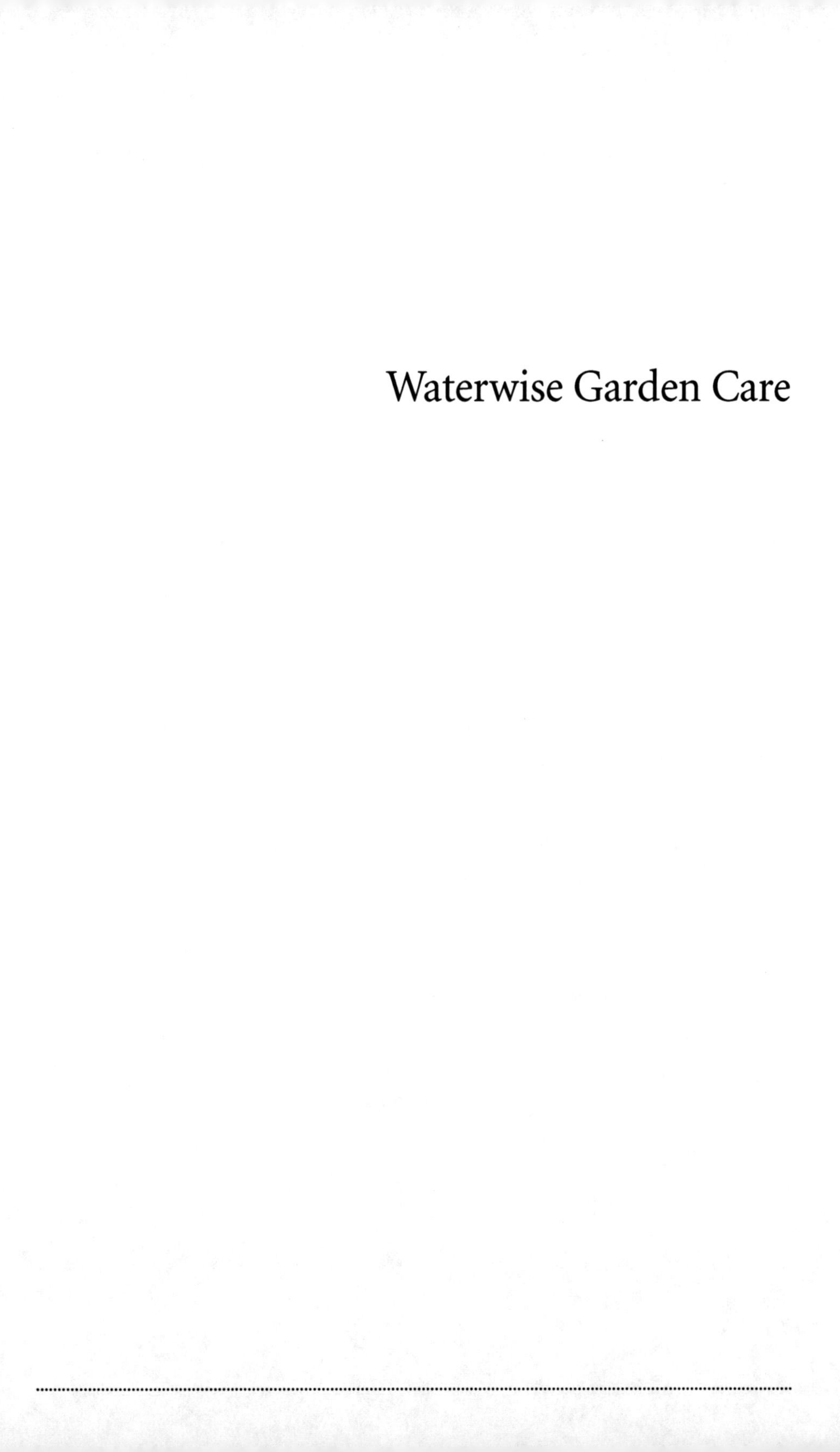

Waterwise Garden Care

Waterwise Garden Care

Your Practical Guide

David Salman, *Chief Horticulturist*

Cindy Bellinger, *Editor*

HIGH COUNTRY GARDENS® PUBLICATIONS
A Division of Santa Fe Greenhouses, Inc.
Specializing in Xeric Gardening

Better Gardening. Naturally!

Published by High Country Gardens Publications,
a division of Santa Fe Greenhouses, Inc.

Printed in the United States

ISBN: 0-9764923-0-X

Book Design by John Cole, Santa Fe, NM

Dedication

This book is for everyone—
especially those who love getting their
hands dirty and making beautiful gardens.

Acknowledgements

This book became a group project with input from some very knowledgeable people. Under Ava Salman, project leader, Dave Abernathy, Jeff Clark, Lacy Gage, Steve Hanchett, Katherine O'Brien, Robert Ross and Mary Ann Walz all contributed specific information about plants and their care. Brent Jarrett helped with editing. Charles Mann took the cover photograph and William Rotsaert did the illustrations.

Hummingbird sipping Penstemon nectar

Amur Maple

Table of Contents

Maiden Hair Grass

Introduction

by David Salman,
Chief Horticulturist

What is Waterwise Gardening?

BESIDES GOOD soil and healthy plants, gardening is all about water. Nothing can live without it; and as the West and other regions of the United States entered a prolonged drought, the face of gardening radically changed.

No longer can we let the hose run as long as we want, or even forget to turn the water off. Our diminishing water supply is asking that we now be conscious of our water use. We need to pay more attention and conserve water at every level of consumption.

So how do you do this and at the same time preserve and create beautiful gardens?

Santa Fe Greenhouses, Inc. began in 1983, and we launched our catalog *High Country Gardens* in 1993. We're located in the arid Southwest and when the drought hit, it hit us hard. Gardening in dry climates can be tricky in the best of times, but some wonderful gardens have existed here for a long time. What we've done is highlight the 'ground level' practices within those gardens, and making those practices available for everyone has become our focus.

Gardening with less water means being waterwise. It means using regionally appropriate native and adapted plants to create beautiful, low maintenance landscapes. It means choosing the right plant for the right microclimate in your yard. It means keeping the soil happy using natural and organic ingredients. It means using mulches to keep precious soil moisture from evaporating. Waterwise gardening is not rocket science; it's simply learning some new common sense techniques to create successful gardens.

Perhaps the heart of waterwise gardening began in 1981 when an environmental planner with the Denver Water Department coined the term *xeriscape.* Taken from the Greek prefix, *xeri* means "dry with little water;" and the goal of this type of landscape is to replace water-thirsty plants with those that are more tolerant of drier growing conditions.

For years we used the word *xeric,* and still do to denote plant characteristics, but some misconceptions have developed. *Xeriscaping* is not 'zeroscaping,' where plants and lawns are torn out and replaced with gravel and little else. Xeriscaping doesn't prohibit the use of thirstier plants, but it does request that gardeners limit their use to the small "oasis" zone of the yard.

Also, because *xeric* means drought tolerant, it's often thought that the plants don't need additional water after they're planted. Drought tolerant means the plant can survive with less water once established. So water a little more the first season or two to establish deep roots. From that point forward xeric plants will be happy with less water.

Waterwise is a better description of the kind of gardening we do now, but the term is interchangeable with xeriscaping. This handbook has the essential facts and techniques you'll need to garden beautifully with less water; and I bet it'll be the most fun you've had gardening in a long time.

Howell's Dwarf Conifer

The Eight Principles of Xeriscaping

1. Plan and Design

Whether you're starting from scratch, or renovating an existing landscape, take the time to plan your design before you start to plant. Create different water use zones and allocate the water where it will most directly contribute to the beauty and comfort of your home.

2. Create Practical Turf Areas

Limit the size of lawn areas and use native grasses as much as possible. These are an excellent drought-tolerant alternative to thirsty Kentucky Blue Grass.

3. Use Appropriate Plants

Use xeric plants for hot, dry south and west-facing areas. Use plants that like more moisture along north and east-facing slopes and walls. Don't mix plants with high and low-watering needs in the same planting area.

4. Improve the Soil

Add organic matter in the form of compost whenever you plant. This helps the soil hold extra moisture.

5. Use Mulches

By covering the soil's surface with some type of mulch, you help retain valuable soil moisture. Mulching also helps capture rainwater by allowing hard rains to soak into the soil instead of running off into the street and drainage areas.

6. Irrigate Efficiently

Don't overwater. Use soaker hoses and drip irrigation to water deeply and encourage deep root growth.

7. Capture Rain and Snow Run Off

Use rain barrels and cisterns to capture water draining off roofs. Run-off from paved areas can be directed back onto the landscape to water trees, shrubs and flower beds.

8. Maintain Your Landscape and Garden Properly

Keep irrigation systems running properly. Avoid the lush, thirsty plant growth that results from over-fertilizing.

earth foods
feeds the soil that
feeds your plants

Fine-tuning Your Xeric Garden

NO MATTER where you live, the basics of gardening don't change. You need fertile soil, good water and hardy plants. However, here in the Southwest some of those basics needed to be refined. Our special conditions here require a more attentive approach. Our soils aren't fertile; they need amending. Water is becoming scarce throughout our region, so certain prudent measures and practices for water conservation need to be put into place. Following are strategies for fine-tuning the waterwise garden.

Soil Preparation

Good soil preparation is essential to gardening success in any climate, but Western soils are typically poor in humus (organic matter) and are moderate, to highly alkaline, and low in phosphorus. Amending soils with organic matter, usually in the form of compost, is necessary for healthy, successful gardens.

We encourage Western gardeners to enrich garden soil through the use of compost and other natural soil minerals and organic amendments. Compost is the single most important soil amendment. It adds much needed humus and beneficial soil microorganisms to the soil. These microorganisms are essential to breaking down organic materials into usable nutrients for plants to use.

Compost also helps sandy soils hold water and improves drainage in tight soils by encouraging microbial activity that creates more pore space, the space between soil particles.

Types of Soil

Basically there are four types of soil:

- ***Sand*** has particles that appear and feel gritty. The space between them is big, so it keeps the soil open and provides good drainage and room for oxygen.

- ***Clay*** has particles that are tiny and dense, and they feel slippery when wet. Clay holds water and nutrients a long time and tends to turn very hard when dry.

- ***Silt*** is somewhere between sand and clay. It's not as tight as clay and not as loose as sand.

- ***Loam*** is a varying combination of all three of the above, mostly silt and sand, with less clay.

Soil Drainage

Because they like drier conditions, xeric plants require excellent soil drainage. Many drainage problems occur due to a high percentage of clay in the soil, or caliche, a white layer of soil cemented together by lime.

When preparing soil, the goal is creating a loose substance that retains the appropriate amount of moisture but will also provide sufficient pore space so plant roots get enough air.

Soil amendments that will improve soil drainage:

- ***Organic matter*** continually decomposes, so more must be added on a regular basis. Use in a ratio of 1 part compost to 2 parts existing garden soil.

- ***Coarse sand*** helps break up clay clumps; don't use fine sand as the mixture will result in a concrete-like consistency. Sand must be mixed generously when amending clay. One part clay, one part sand is recommended. Testing percolation time (how quickly water moves through the soil) will verify if there is enough sand in the mix.

- ***Gypsum*** will help break down clay particles that lump together, thus improving the circulation of air and water. Add generously (at recommended rates) when preparing new planting areas.

- ***Greensand*** also helps break down the hard texture of clay. Add when preparing new planting areas.

- ***Gardening Tip:*** The use of gypsum and Greensand will sometimes take several years to effect a change. Annual top-dressing with these ingredients each fall helps facilitate drainage.

- ***Gardening Tip:*** Poorly drained soils often requires special treatment to improve the drainage for many xeric plants. A good solution is mixing clay with plenty of coarse sand and a moderate amount of compost, then filling raised planting beds or creating mounded ("bermed") beds.

Soil Recipes

Different plants require different soil compositions. The following list is a general guideline for the kind of soil that certain plants like:

- **Hardy Perennial Flower and Bulb Beds (plants that prefer moisture-retentive soils with high humus content)**
  Spread and dig in to a depth of 12 to 15 inches, Soil Mender® Blend (a mixture of well composted cotton burrs and cow manure), Planters II trace mineral fertilizer, Yum Yum® Mix fertilizer and Superphosphate (or soft rock phosphate). In sites with heavy clay and "caliche" (white chalky subsoil), dig in ample gypsum and Greensand and extra Soil Mender® Blend deep into the subsoil.

- **Vegetables and Annuals**
  Same as above, but in the fall or spring dig in more Soil Mender® Compost to replenish the nitrogen level in the soil every year. Annuals and vegetables pull a lot of nitrogen from the soil. Planters II can be added every other year.

- **Xeric Plants (including penstemon and cold-hardy cacti and succulents)**
  Add Planters II and Yum Yum® Mix by the handful into individual planting holes or at recommended rates when preparing a bed. Use only a handful of Soil Mender® Blend in each planting hole if the soil is hard or heavy clay.

- **Topsoil for Raised Beds and Planter Boxes**
  3 parts clean, screened soil plus 1 part Soil Mender® Blend, in addition to Planters II (at 10 lbs. per 1 cu. yard) and soft rock phosphate (at 4 lbs. per 1 cu. yard)

- **Potted Annuals and Herbs**
  4 parts bagged potting soil, 1 part Soil Mender® Blend, a few handfuls of Yum Yum® Mix or slow release Gro-Power fertilizer, and water-holding granules.

- **Planting Trees and Shrubs**
  Same as for flowerbeds, but also work sulfur into the soil for plants that are sensitive to alkaline soil such as roses, ornamental and fruiting peaches, plums, cherries, aspen, wisteria, quince, and Japanese maple.

- **Blue Grass/Fescue/Thyme Lawns**
  Same as for flower beds, but also rototill Soil Mender® Blend and other ingredients to a depth of 6 inches.

- **Legacy® Buffalo/Blue Grama grass**
  Planters II, and 1/2 cu. yard of Soil Mender® Blend per 100-sq. feet (or Gro-Power 5-3-1 fertilizer at 5 lbs. per 100-sq. feet).

Products for Soil Improvement

Following is a list of the different recommended products for soil amendments:

- ***Yum Yum® Mix*** is an organic soil-enhancing fertilizer well-balanced with alfalfa, cottonseed and kelp meals; it also has Green-sand, trace minerals and phosphate. The mix has a low (N)itrogen, (P)hosphorous and (K)potasium formula that is good year round; won't burn plants.

- ***Soil Mender® Blend*** is a pre-mixed blend of organic composted cow manure and cotton burrs. Just pour it from the bag and mix it into the soil as you plant.

- ***Soil Mender® Mulch*** is a coarse textured organic cotton burr compost. Cotton burrs are excellent for composting because of their naturally high nitrogen levels.

- ***Soil Mender® Compost*** is a rich, fine textured organic composted cow manure mixture. It's recommended for digging into the soil for vegetable gardens and ornamental plantings. A little goes a long way as this compost has high levels of NPK.

Sources for Soil Nutrients

- ***Soft Rock Phosphate*** is a natural source of phosphorous, an essential plant nutrient, lacking in Western soils that is needed by plants for root growth, flowering and fruiting.

- ***Soil Sulfur*** is a natural long lasting soil acidifier used for plants that require less alkaline soil conditions.

- ***Greensand*** is a natural iron and potassium source that helps fortify plants suffering from iron chlorosis, a deficiency that causes sickly yellow foliage. Scratch into the soil in the fall and early spring. Dug in, it also helps break up clay soil.

- ***Planters II*** is a granular and non-burning natural trace mineral fertilizer that contains over 30 trace minerals that are often low or lacking in many soils. Adding this will result in robust rooting and flowering, especially in native and adapted xeric species that naturally grow in mineral soils.

Soil Mender® Compost and YumYum Mix®

Watering

No matter what kind of garden you have—formal, informal, ornamental or vegetable—and no matter what region you live in, having a garden is all about watering. But today with the growing concerns of drought and water shortages the biggest gardening questions relate to watering: how much and when?

Types of Water and Quality

Harvested Water

The recent droughts have made us more aware of the lack of water, and more people are installing water catchment systems—from complicated cisterns to simple buckets under roof gutters. Captured rain and snow melt is free water, and perhaps the best water. After all, doesn't your garden look happiest after a good rain? Water straight from the sky (or captured through down spouts) is neutral and soft. There's nothing like rainwater to perk up drooping plants.

- ***Gardening Tip:*** Even a light rain can yield hundreds of gallons of water from roof catchment systems.

Well Water

Well water is a common source for both city and country residents. But whether from community or private wells, this water in much of the western United States is hard and alkaline. Plants appreciate the moisture, but it's not their favorite kind. Also, during severe droughts restrictions are often placed on use of well water so as not to deplete the wells.

Grey Water

In recent years, using grey water on landscapes has become a popular topic. This is water from dishwashers, bathtubs and washing machines. Because soaps, body oils and bacteria haven't been documented to be harmful, this water is getting more attention for garden use. Most concerns focus on using grey water in vegetable gardens because those plants will be eaten.

New or modified grey water systems are expensive, but catching grey water in buckets under sinks and in showers is easy. The conservation angle of grey water is quickly catching on, but use restrictions depend on local ordinances.

Types of Watering Systems

By Hand

This is the trusty, good old-fashioned way of watering—holding a garden hose to each plant or letting the hose run in selected beds. There are hose attachments that help with certain situations: sprinklers, soakers. Also hard-to-reach plants can be watered by using a watering can.

Drip Irrigation

This type of watering system is a wonderful time saver, but can be mechanically challenging. Tubing is hooked to a water source on a timer and pressure reducer that directs water to a system of black plastic tubing that snakes through the garden. Small spaghetti-sized emitters attach to the tubing and deliver water to your plants. Some emitters allow for 1 gallon per hour; some allow for 2 gallons per hour. This system can be adapted to different watering needs.

The beauty of this system is that they grow as your garden does by adding more tubing and emitters. Usually one emitter per plant will do with perennials and annuals, but larger woody plants will need more, depending on the plants' water needs.

The downside of drip irrigation is the maintenance. The emitters can get plugged with dirt or pop out of the main hose. Sometimes the piping is chewed by gophers, drying out entire garden areas. But for the most part drip irrigation systems deliver water efficiently to the root zones.

Many existing sprinkler systems can be retrofitted to accommodate multi-line drip emitters. Do-it-yourself drip irrigation kits are also available to run efficient drip systems off existing hose faucets.

Soaker Hoses

These are porous hoses that lie on top of the soil (often covered with mulch to hide them). Water gently seeps out of the hose along their whole length. The use of soaker hoses or 1/4 inch diameter tubing with emitters is a very efficient method of watering flowerbeds. This is also a good way to water closely spaced plants.

Root Feeders

These devices push into the ground with a hose attached for deep underground watering; good for trees and shrubs.

Watering Requirements for New Plantings

Whenever possible, plant when the weather is not too hot; fall and spring plantings require less frequent irrigation.

- Some native xeric shrubs and trees will establish on a 1-day per-week schedule. As a general rule, a 5-gallon potted xeric shrub or tree needs 5 gallons of water applied 1 time per week.

- Most trees and shrubs and xeric perennials will become established on a 2-days per-week schedule.

- Most perennials will need 1 to 2 gallons of water applied 2 times per week. Trees and evergreens will need approximately 20 gallons per week. Sandy soils may require extra amounts of water.

Making New Plantings Water Efficient

- Some gardeners like to use water-holding compounds that are added to the soil at planting time, which can cut watering needs up to one half.

- Fortify the roots with a root stimulator to encourage strong root growth and minimize transplanting shock. Water with the root stimulator every 1 to 2 weeks for at least 3 applications.

- Mulch all your plantings to a depth of 2 inches.

- Create a "well" for water around newly installed plants with a circle of plastic edging, a ring of rocks or a ring cut from the top of a plastic nursery pot. This is highly preferable over a ring of soil, which melts and holds less and less water over time. Fill the "well" twice when watering to completely saturate the root ball. Not needed when putting plants on drip systems.

- Use saucers under pots, hanging baskets and window boxes. This allows the containers to re-absorb water that runs out the bottom.

- When using drip systems, be sure at least one emitter is placed directly over the root ball. Other emitters should be placed off to the sides to moisten the surrounding soil. When running the system to water new plantings, apply 5 gallons of water per 5-gallon sized plant, 10 gallons of water per 10-gallon sized plant.

- When planting trees and shrubs, holes dug in advance can be filled with water and allowed to drain. (Don't wet the backfill soil as this will cause it to compact when re-filling the hole.) This subsurface water does not evaporate from the soil and will be available to deep growing roots.

Water Maintenance for Established Landscapes

A watering schedule depends on where you live, the type of soil in your garden and how much precipitation occurs during the growing season. In general, non-xeric plants need irrigating 1 to 2 times weekly in summer. Xeric plants need a deep soaking once every week to ten days. Very xeric trees and shrubs need a deep soaking monthly if conditions are dry.

- Watch plants for sagging foliage and a grayish color; this is their first cry for water.
- Water early in the morning during the warm months to lessen evaporation.
- Deliver water slowly so it can penetrate and not run off.
- The goal is to keep root systems from completely drying out.
- Pay attention to natural weather conditions. Don't water when it is windy. Wind quickly evaporates water.
- Keep your garden mulched; this lessens evaporation.
- Set drip system timers to thoroughly soak the soil once a week instead of running it for a short duration many times per week. Established trees and shrubs will need one 3-to-4 hour soaking per week. Perennials will need one 2-hour soaking per week.
- Water established and mature trees/shrubs once every 2 to 3 weeks. Use drip irrigation or soaker hoses covered by a layer of mulch to apply water most efficiently.

- Kentucky Blue/Fescue turf grasses will struggle with a once per week irrigation. It helps to use a wetting agent sprayed over the lawn to improve water penetration. Then top-dress with 1/2 inch layer of Soil Mender® Compost to mulch the roots.

- ***Gardening Tip:*** Non-native Kentucky Blue Grass lawns use up to 75% of your landscape's total water.

- Remember to water trees that are planted in lawns, even if you stop watering the lawn. Place sprinklers to water out to the drip line of the tree while leaving open areas of turf un-irrigated. Established trees need a soaking once every two weeks.

- Use organic fertilizers that fortify the soil by adding organic matter, which increases its water-holding capacity.

- Add them to your irrigation water once a month through the summer and fall.

- Avoid fertilizing with chemically derived fertilizers. Continual use of chemical fertilizers decreases the soil's ability to hold water by destroying the spongy texture created by humus and beneficial organisms.

Collecting water in a rain rarrel

Mulch

Mulch is any substance spread over a garden to form a protective layer. In late spring through fall, mulch keeps the soil from drying out. During the winter and early spring, mulch keeps the ground frozen, which helps prevent roots from breaking due to frost heaving.

The tiny root hairs that grow off a plant's main root are responsible for getting water and nutrients to the plant. For root hairs to live they must stay moist. If they dry out, they die. Even if the main root system survives the winter, the plant will suffer if the root hairs are gone. One way to ensure that roots don't dry out is to use mulch.

Types of Mulch

- ***Organic*** includes bark, wood chips, rotted sawdust, straw, leaves, pine needles, newspaper and cardboard—anything that decomposes.

- ***Inorganic*** includes stone, gravel and woven geotextiles ("weed barriers").

All Mulches Help

- Conserve soil moisture
- Keep down weeds
- Reduce erosion and help capture rain water
- Keep plant roots cool during the summer
- Provide insulation during the winter
- Make gardens attractive

How to Mulch

- Remove any existing weeds in bedding area.

- Spread at least 1 to 2 inches of mulch in flowerbeds; and 2 to 3 inches around shrubs and trees.

- Place over root zones; never mound over crowns or near trunks, as this might cause wet rot.

- Blanket the area to a uniform depth. Any remaining low or bare spots are prone to weed problems.

Other Mulching Tips

- For trees, extend a ring of mulch from the trunk out to the drip line. Depending on the tree size, this ring can be up to 6 feet wide.

- Xeric plants prefer a non-packing mulch that won't stay wet and hold moisture around the crown or base of the plant. Gravel and pine needles are best.

- As organic mulches decompose, they become part of the soil and need to be replenished every year.

- Don't let organic mulch build to depths greater than 4 inches; this will smother the soil and deprive roots of oxygen.

Fertilizer

This is a nutrient source, either liquid or dry, that is applied to the soil or sprayed onto the foliage.

Types

- ***Organic fertilizers*** are made from decomposed plant matter or manures that supply the soil with major and minor nutrients and trace minerals. Once they decompose in the soil their nutrients are available for up-take by plant roots.

- ***Inorganic fertilizers*** are made up of chemicals that do provide plants with needed nutrients, but they don't add humus to the soil. Prolonged use of chemicals makes soil sterile, making plants dependent on continuous applications.

When to Apply

The growing cycle of plants not only determines when to fertilize but what kind of fertilizers need to be used. And keep away from chemical derived fertilizers as they increase health hazards for both humans, animals and insects.

- ***Fall*** is the preferred time to fertilize perennials, shrubs, trees and grasses. Fertilizers high in potassium and phosphorus encourage strong root growth. The best time for fertilizing in the fall is after a few light frosts or when the trees begin dropping their leaves. Fall fertilization also promotes recovery of root systems damaged by drought.

- ***Spring*** is also a good season to fertilize. Needled evergreens (conifers) can be pushed by nitrogen rich composts or fertilizers. Apply a few weeks before their spring growth spurt.

- ***Gardening Tip:*** During the summer months plants (such as aspen, purple leaf plum, peaches and wisteria) can exhibit chlorosis or yellow foliage. This is a sign of iron deficiency. Fertilize with Greensand, an all-natural iron source that helps "green-up" the plants next summer. For an immediate "green-up," chelated iron formulations can be sprayed directly on the foliage.

How to Read Fertilizer Labels

Printed on all bags and bottles of fertilizers are three hyphenated numbers such as 2-3-2. This is the NPK ratio with letters standing for:

N = nitrogen (stimulates foliage)
P = phosphorus (encourages strong root growth)
K = potassium also known as potash (promotes hardiness)

- ***Gardening Tip:*** Fertilizing in the summer with high nitrogen formulations will result in rapid, thirsty new growth that is much more susceptible to injury from drought. Root systems can also be damaged and made susceptible to soil-borne diseases.

How to Apply Fertilizers

- ***Liquid*** usually comes in concentrated forms and needs to be mixed with water. Follow the directions on the bottle for dilution mixtures. Then either pour directly on the soil under the plant or spray onto the leaves, as with many indoor plants.

- ***Granular*** is dry and grainy and raked into the soil either with fingers, a small rake or hand cultivator.

Stimulating Root Growth

All new transplants depend on vigorous new root growth to thrive, and using a root stimulator at the time of planting reduces shock.

- Use a mixture of 1 teaspoon Saltwater Farms liquid seaweed fertilizer and 1/4 teaspoon Superthrive per gallon of water. Water initially with clear water; then apply the root stimulator mix, saturating the root ball and surrounding soil. Re-apply this mixture 3 or 4 times at two-week intervals for optimum results.

- Liquid seaweed can be sprayed directly on the foliage to help reduce transplant shock. This is especially helpful if lots of rain has saturated the soil, making root application of the Superthrive/seaweed mixture impractical.

Notes for Soil Preparation, Watering, Mulch and Fertilizer

Controlling Insects and Critters—Naturally

Insects

THE BEST way to prevent insects from invading your garden is to keep a good harmonious balance between soil, water, fertilizer and mulch. When all these elements work together, a garden is happy. What we've learned here in the Southwest the past few years with our massive piñon tree die-out is stressed plants become weakened. In turn, weakened plants become susceptible to disease, and diseased plants then become easy targets for insect infestations.

So how can this idea apply to our gardens? Same thing—keep them moist, fed and happy. It's the best way to keep the intricate balances working all the time.

A partial list of beneficial insects

- ***Ladybug***—many species of this tiny beetle have an enormous appetite for aphids—one of our most common plant pests. Others prefer scale insects and mites and are very effective in reducing infestations.

- ***Praying mantis***—it looks like a leaf and the disguise allows the mantis to grab more harmful insects.

- ***Lacewing***—the larvae is an enemy of small caterpillars, aphids, and other soft-bodied insects.

- ***Ground beetle***—common under logs and debris, feeds on a variety of insects.

- ***Wasp***—feed on other insects.

- *Gardening Tip:* If you've enticed the good insects and the nasty ones persist, the next step is using traps, barriers and hand-picking. Organically derived chemicals (such as red pepper) should be used only as a last resort because they will kill beneficial insects as well as the pests. Insecticidal soaps can be used as well, but may burn foliage while attacking the insects.

Dealing with the Ips Engraver Beetle, commonly called the Bark Beetle

These small beetles, 1/8 to 1/4 inch long, reddish brown to black, with a pronounced cavity in the rear, fly to the trunk and main branches of trees stressed by drought and previous beetle attacks. This beetle mostly affects the piñon tree; its relatives bore into ponderosa and juniper.

Symptoms of Infestation

- Treetops start dying back quickly and uniformly.
- Sawdust is sometimes found at the base of the tree and on tops of branches where they attach to the trunk.
- Where beetles chew into trees, they leave behind "pitch tubes," or conical mixtures of sap and sawdust. These "pitch tubes" are attached to the tree bark and are reddish-orange when fresh.

Beetle Cycle

- The Ips Beetle is generally active April through September when the weather is warm.
- The beetles bore into the bark and construct egg galleries, pushing the boring dust out through the entrance holes as they continue boring underneath the bark. When eggs hatch into the larvae stage, the larvae continue creating galleries, further damaging the flow of fluids in the tree.
- The Ips Beetle usually carries the Blue Stain Fungus, which is also deadly to trees. Consequently, a tree that hasn't died from the bark beetle attack will likely die from the fungus.

Prevention and Treatment

- Spraying cannot rid infested trees of the beetle or bring the tree back to life.

- Spraying can help prevent attack, but because new products are constantly being developed, we recommend contacting your County Extension Service for the latest advice.

- The most effective preventative measure to reduce the possibility of attack is to keep trees regularly watered. Irrigate piñon trees every 2 to 3 weeks in the summer, and monthly in the winter. Mulching to the drip line will retrain soil moisture and gradually increase the soil's humus content. This provides trees with nutrients and increases the soil's water-holding capacity.

Critters

Moles and Gophers

One flower that these critters don't like is the daffodil. For a more direct approach, try treating the soil with Mole and Gopher Med, a natural repellent derived from the Castor bean. We also recommend Bulb Guard when planting spring blooming bulbs.

Deer and Rabbits

If hungry enough, these browsers will eat just about anything. One suggestion is planting a crop of their favorite munchies away from your garden just for the animals.

Below is a chart of plants that deer and rabbits typically don't eat.

	Rabbit Resistant	Deer Resistant
Achillea (Yarrow)	X	X
Agave (Century Plant)	X	X
Aconitum (Monkshood)		X
Agastache (Hyssop or Hummingbird Mint)	X	
Agastache cana (Texas Hummingbird Mint)	X	X
Alcea rosea (Hollyhock)		X
Allium (Ornamental Chives)		X

	Rabbit Resistant	Deer Resistant
Allysum saxatile and Annual Allysum		X
Amorpha canescens (Leadplant)		X
Anemones		X
Anethum graveolens (Dill)		X
Antirrhinum (Snapdragon)		X
Aquilegia (Columbine)	X	
Artemisia	X	X
Astilbe		X
Aubrietia (Rock Cress)		X
Berlanderia (Chocolate Flower)		X
Cacti	X	X
Caryopteris (Blue Mist Spirea)	X	X
Calylophus hartwegii (Sundrops)	X	
Cerastium tomentosum (Snow in Summer)		X
Chamaebatiera millifolium (Fernbush)		X
Chrysothamnus (Chamisa)	X	X
Consolida ambigua (Larkspur)		X
Coreopsis		X
Cytisus purgens (Spanish Broom)	X	X
Delphinium		X
Digitalis (Foxglove)	X	X
Dicentra spectabilis (Bleeding Heart)		X
Echinacea (Coneflower)		X
Eriogonum umbellatum (Buckwheat)	X	
Erodium (Stork's Bill)	X	
Falugia (Apache Plume)	X	X
Gaillardia (Blanket Flower)	X	X
Geraniums (Hardy Perennial Geraniums)	X	X
Helianthus (Sunflower)		X
Hemerocallis (Daylily)		X
Hesperis matronalis (Dame's Rocket)		X
Iris		X
Lamium maculatum (Dead Nettle)		X
Lavandula (Lavender)	X	X

	Rabbit Resistant	Deer Resistant
Liatris punctata (Gayfeathr)		X
Linaria vulgare (Toadflax)		X
Lychnis coronaria (Rose Campion)		X
Mahonia fendleri (Mahonia)	X	
Menta (Mint)		X
Narcissus (Daffodils)	X	X
Nepeta (Catmint)	X	
Oenothera missouriensis (Evening Primrose)		X
Origanum (Oregano)	X	X
Oxytropis (Locoweed)	X	
Penstemons	X	
Peony	X	
Petalostemon purpurea (Prairie Clover)		X
Perovskia (Russian Sage)	X	X
Philadelphus lewisii (Mock Orange)		X
Potentila fruticosa (Potentilla)		X
Ratibida (Prairie Coneflower)		X
Rosmarinus (Rosemary)	X	
Rudbeckia (Black Eyed Susan)		X
Salvia (Sage)	X	
Santolina	X	
Scorphularia macrantha (Redbirds in a Tree)	X	
Sedum (Stone Crop)	X	
Silene laciniata (Mexican Catchfly)	X	X
Solidago (Goldenrod)		X
Sphaeralcea munroana		X
Stachys byzantina (Lamb's Ears)		X
Stachys coccinea (Hedgenettle)	X	X
Stanleya pinnata (Prince's Plume)	X	
Tagetes (Perennial Marigold)	X	
Tanacetum vulgaris "Crispum" (Tansy)	X	
Thymus (Thyme)	X	X
Kniphofia uvaria (Red-Hot Poker)	X	
Vinca major		X
Yucca	X	
Zauschneria species (Hummingbird Trumpet)		X
Zinnia grandiflora	X	X
Zizophora clinopodioides (Blue Mint Bush)	X	X

Notes for Controlling Insects and Critters—Naturally

Angus
Tyme in flagstone

Guide to Planting and Caring for Plants

THE DIFFERENT types of plants all have special ways they like to be planted. Following are guidelines for planting:

- **Bulbs**
- **Container gardens**
- **Native grass lawns**
- **Perennials and ornamental grasses**
- **Thyme lawns**
- **Trees and shrubs**
- **Wildflower seeds**
- **Winter-hardy succulents and cacti**

- ***Gardening Tip:*** Group plants together by their cultural needs; that is, the same growing requirements. For instance, plants that like drier conditions will not do well with those that need frequent watering. Likewise, sun lovers don't do well next to shade loving species.

  Cultural elements to keep in mind:

  - Sun and shade
  - Watering
  - Soil fertility
  - Soil drainage

- ***Gardening Tip:*** Plants also prefer different types of mulch. For example lavenders, cacti, diascia, and penstemons like gravel, but some of the more water loving plants (usually those that like richer soil) want organic mulch.

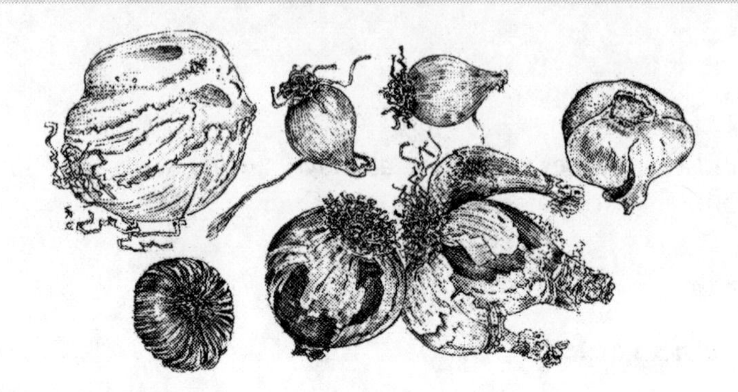

Planting Bulbs

Bulbs are little globules of wonder. Hardened shells with soft inner bodies, they bloom and offer a delight of color to any garden.

How to Purchase

- Bulbs should be firm. If soft, they have internal rot and are not healthy.
- Choose bulbs with minimal leaf sprouts.
- Choose the bigger bulbs. Bulbs are graded by circumference. Smaller sizes are less expensive, but often produce inferior, undersized plants with fewer flowers. They will take several growing seasons to catch-up with larger bulbs.

When to Plant Different Types

Fall blooming bulbs—plant between late August and early October.

- Sternbergia lutea (Lilies of the Field)
- Fall Blooming Crocus—many different varieties

Spring and summer blooming bulbs—plant from early September to late November, or until the soil freezes. Feeder roots must grow before freezing weather arrives to ensure healthy blooms in the spring. A good rule of thumb is to plant them about 6 weeks before the ground is frozen in your area.

Early bloomers—

- Chionadoxa
- Crocus and miniature Iris
- Galanthus elwesii (Snow Drop)
- Wildflower Tulips
- Muscari, more commonly known as Grape Hyacinth
- Scilla
- Daffodils, regular and miniature

Mid-spring bloomers—

- Daffodils, regular and miniature
- Tulips

Late-spring bloomers—

- Daffodils
- Darwin hybrid tulips
- Bearded Iris

Early-summer bloomers

- Allium species Ornamental

Soil and Site Requirements

- Well-drained soil enriched with Soil Mender® Blend.
- Many bulbs such as Tulips and Alliums prefer a full sun exposure. Muscari, Daffodils, Galanthus, Hyacinthoides will tolerate partial shade.

How to Plant and Water

- Either using a trowel or bulb digger, make shallow holes for the smaller bulbs and deeper holes for the larger bulbs. Tulips, Daffodils and ornamental Alliums do best planted at least 8 to 10 inches deep. If your soil is very shallow or rocky, you can plant bulbs less deeply, but they may not survive as long.
- Before placing bulbs in hole, add a small amount of soft rock phosphate and Yum Yum® Mix fertilizer. Cover bulb completely with soil.
- After planting, add a top-dressing of compost or other organic material and water-in thoroughly.
- Water on a regular basis, once every three weeks to a month, when the soil isn't solidly frozen.
- Continue to water through the spring as the bulbs begin poking out of the ground. This will insure longer lasting blooms.

Care Following Blooming

- Allow foliage to brown and fade naturally. Because the leaves are needed to feed the bulb in the ground, premature removal of green foliage weakens the bulb and leads to fewer blooms the following year.
- Additional top-dressings of organic fertilizer and mulch every fall feed the soil and insure that bulbs multiply strongly and bloom heavily next spring.

- Treat the soil of non-daffodil species with Mole and Gopher Med in late summer to repel bulb-eating gophers, moles and voles.

Designing with Bulbs

- For a range of continued bloom throughout the spring, keep in mind the bloom time. Be sure to plant early, mid- and late-spring blooming combinations.

- Use enough of a single color and variety to make a definite impact; one bulb won't give a very strong presence on its own. Bulbs usually look best placed in clumps of 5 to 7 of a single variety.

- Drifts (groups) of closely planted bulbs in a perennial border create a blooming carpet while the perennials are just starting their foliage. By the time the bulbs have stopped blooming, the perennials will be tall enough to hide the dying foliage of the bulbs.

- Plant smaller growing types into carpets of early blooming groundcovers like Veronica and 'Pink Chintz' Thyme. Daffodils combine beautifully with spreading Symphytum (ornamental comfrey).

- Plant taller bulbs into daylilies and other tall late-spring and summer blooming perennials so their bulb foliage gets covered as it goes dormant.

- Space bulbs somewhat randomly when planting. Bulbs planted in rows look artificial.

Planting Container Gardens

The endless and attractive combinations of flowers, foliage and edible plants explain the rising interest in container gardens. Also, container gardens are the most water efficient way to grow annuals and vegetables. Container garders are naturally water-thrifty because all the water applied is contained in the pot.

Choose the Right Kind of Pot

- Plastic, foam and glazed clay pots all retain water better than unglazed terracotta or wood.

- Make sure the container has adequate drainage. Although you want to minimize water loss, you also don't want to create a wetland bog. If left in saturated soil, roots will suffocate and rot.

Choose the Right Size of Pot

- The larger the container, the more resources are available to the plants—water and soil nutrients.

- The more soil a container holds, the less frequently you'll need to water it.

Use the Proper Soil

- Use a soil with good water-holding capacity. A bagged, soil-less potting mix is a good choice.

- Avoid using soil from your garden beds. It compacts in pots and becomes unusable for the plants. Similarly, using bagged topsoil is also unsatisfactory unless mixed heavily with a soil-less potting mix.

- ***Gardening Tip:*** To mix your own potting soil follow this recipe:
  For each 2.5 – 3 cubic foot bag of potting mix add:
  • 2/3 cup water-holding crystals
  • 3/4 cup Yum Yum® Mix fertilizer

  The water crystals act as tiny sponges, soaking up water and holding it. Yum Yum® Mix is an organic fertilizer that slowly releases balanced nutrients to the plants, lengthening bloom-time and promoting healthy growth.

Choose the Right Plants

Not all plants are well-suited to containers. Some are simply too big, others grow too slowly to fill out in one season. Nonetheless, there are many that perform splendidly in pots. Here are a few suggestions that are well-suited to containers and can take hot, dry conditions:

- ***Annuals & Cold Tender Perennials:*** African Daisy, Dahlberg Daisy, Gazania, Lion's Ear, Marguerite Daisies, Marigolds, Portulaca, *Salvia*, Sanvitalia, Society Garlic, Verbenas, Vinca, Zinnia.

- ***Perennials*****:** Artemisia, Cascading Oregano, Coreopsis, Creeping Thymes, Festuca Grasses, Gallardia, Gaura, Lavenders, Mexican Hat, Penstemon, Salvia, Santolina, Sedum, Stipa Grass, Veronica, Vinca Vine, Yarrows.

- ***Herbs:*** Lavender ('Goodwin Creek,' 'Tenerife,' Spanish Butterfly), Oregano (Italian & Greek), Rosemary (trailing 'Irene' & upright 'Arp,' 'Spice Island,' 'Tuscan Blue'), Sage (Garden, Golden, Purple & Tri-Color), Thyme (English, 'Ray Williams' and Silver Leaf).

- ***Vegetables:*** Most veggies aren't drought-tolerant, so container gardening is an excellent way to grow them because they'll require less water than in the ground. Choose dwarf varieties such as Bush Cucumber and 'Patio' Tomato. Runner beans and corn also do great in pots, given enough space. Position pots with vegetables in areas that have afternoon shade.

- ***Cactus & Succulents:*** What better choice for a hot, dry container than a cactus garden? Great choices include native Claret Cup Cactus, Ice Plant (*Delosperma* & *Ruschia*), *Agave*, Yucca and *Dasylirion*.

- ***Trees & Shrubs*:** Consider using dwarf trees, conifers and shrubs in container gardens. Many varieties make excellent additions to combination planters. Remember to either choose slow-growing varieties or true dwarfs. Dwarf conifers such as Mugo Pine 'Slow Mound' give year-round interest to pots.

- ***Grasses*:** Some grasses grow tall, others mound. It all depends on the look you want. Some suggestions include Friber Optics, Purple Fountain, Toffee Twist Sedge, and Ribbon Grass.

Group Plants According to Cultural Needs

Choose plants for each pot that require the same watering, feeding and sun conditions. You don't want one plant that likes a daily watering and one that likes a weekly watering. By choosing plants carefully, you can create a container garden that will last year after year.

- ***Gardening Tip:*** If you want your container garden to winter over, choose plants that are one zone below where you live. Or bury your pots in the ground or mulch and protect with straw.

Arrange Plants According to Height, Color and Texture

The various ways plants can be grouped is an exciting aspect of container gardens. Arrange them by height, color and texture. Consider growing grasses with plants that cascade over the edge or soft flowered plants alongside spiky leafed plants.

Where to Put Your Container Garden

Before planting a container, decide where to place it and perhaps fill it in place. Once filled it could be difficult to move. Containers placed near walls or concrete or flagstone patios will readily absorb the heat. Not all plants will appreciate such harsh conditions, and will need more frequent watering, or moving. Direct sun after 2pm quickly dries out pots. Even sun-loving plants will have enough by this time of the day. Consider placing pots in filtered or dappled shade all day long.

Watering Container Gardens

Your new container garden will use less water than similar plantings in the ground. However, because of the limited water-holding capacity of pots, you'll need to water them more frequently. Depending on the size of the pot, its placement and the plants selected, you may need to water as often as once a day. Plants requiring low water use that are placed in large containers and get afternoon shade may need water only every other day or less.

One key to water retention is mulch. Just as flowerbeds benefit from a good layer of mulch, so can your container garden. Mulch slows evaporative water loss from the surface, and insulates the soil. Moderate soil temperature is also better for roots.

Fertilization

Many container gardens can use additional fertilizer throughout their growing season, especially annuals and vegetables. Use a water-soluble fertilizer, applied once a week, or top-dress with a slow-release fertilizer in late spring. Again, consider grouping your plants according to need.

Maintenance

Many plants will bloom longer and more prolifically if deadheaded regularly. Deadheading is simply the removal of spent blossoms. However, make sure you snip off the whole flower stem, and not just the petals. A lot of plants drop their petals after the flower has been pollinated, leaving the ovary intact. If left alone, the plant will put its energy into producing seeds inside the ovary, and not into more flowers.

Planting Native Grass Lawns

As we become more waterwise in our gardening techniques, the idea of lawns here in the West is changing. No longer is the lush, Kentucky Blue Grass the ideal. Living with droughts, it only makes sense to use native grasses.

Blue Grama Grass

This makes a fine native turf grass when seeded at the proper density. We recommend seeding at a rate of 3 to 4 lbs. per 1000-sq. feet Seeds can be sown beginning in late spring/early summer when night temperatures reach 60°F. Seeding can continue through the summer months until 6 weeks prior to first average frost date. This grass can also be planted from plugs.

Site Preparation

Grama grass is well adapted to poor soils and soil enrichment is not required. However, improving the soil prior to planting will help sandy soils hold more water; and added nutrients will help the grass thicken up more quickly.

- When preparing to plant a non-lawn area with Grama grass seed, clear the area of weeds.

- If the site has an existing lawn and preparations can be done far enough in advance, the existing lawn can be left unwatered until it dies. After a month or more, when the roots have had time to rot, the dead grass can be rototilled into the soil. Rototilling live grass into the soil is not recommended as the turf grass will regrow.

Planting Seeds on a Slope

- When planting sloped areas, plant the seed, then cover the soil with an erosion preventing material, like seed-free wheat or barley straw or aspen excelsior matting that can be pegged into place. A biodegradable soil binder can also be used to hold the soil in place. Contact your County Extension agent for sources of erosion prevention materials.

Preparing the Seed for Planting

- Measure the area to be seeded and calculate the square footage.

- Next, thoroughly mix the seed in a bucket half filled with moist sand. Grama grass seed is very fluffy and this technique helps sow the seed evenly over the entire area. (For example, if you are sowing an area of 500-sq. feet, mix 2 lbs. of seed with the sand and spread the mixture.)

- Measure the next area and weigh out the appropriate amount of seed. Mix the second batch of seed with sand and sow. Planting in sections will prevent over or under-seeding any given area.

Preparing the Soil and Sowing the Seed

- Rototill the area to be seeded to a depth of 3 to 4 inches.

- Take a bow rake and rake up any roots, rocks and soil clods. Then comb the soil with the rake to leave it covered by shallow furrows.

- If you want to enrich the soil, add compost (at the rate of 1/2 cu. yard per 100-sq. feet) or Gro-Power 5-3-1 fertilizer at the rate of 5 lbs. per 1000-sq. feet and rototill into the soil to a depth of 4 to 6 inches.

- Broadcast the seed/sand mixture by hand. Turn the rake over, using the flat side to smooth the soil and cover the seeds.

- Top-dress with a thin layer of mulch to retain moisture and keep birds from eating the seed.

Time Needed for Germination

- Warm night temperatures, evenly moist (but not soggy) soil, and rain showers help Grama grass germinate quickly, usually within 7 to 10 days of planting. Less than ideal conditions such as windy weather, dry soil and cool nights (due to a cold snap) will delay sprouting. If the seeds have not germinated within 15 days, it may be necessary to re-sow more seed.

Watering Newly Germinated Seed

- After sowing the seed, water thoroughly so the soil is wet to a depth of 4 to 6 inches. Be prepared to water twice daily, morning and evening. Each time, water enough to keep the top 1 inch of the soil damp. Continue watering twice daily until the grass has germinated.

- Be sure to check the depth of soil moisture in several spots around the new lawn area after watering to be sure it's wet enough.

- Once the seeds have germinated, irrigate just enough to keep the soil damp (not muddy) to a depth of 2 inches. Initially, depending on each situation, this may require a once daily watering (morning or late afternoon) for a week to 10 days. As the grass begins to grow, watering frequency should be reduced to once every 2 or 3 days. At the end of 4 weeks, the grass should need watering only once a week.

Watering Sloped Areas

- Be sure to mulch. Then water with a fine spray, just enough so the water does not run off heavily. Come back and re-water 2 or 3 more times until the soil is wet to a depth of 2 inches.

Maintenance

- ***Watering*:** Once established, Grama grass is very drought tolerant. To stay green and grow actively, it may need extra water during the hottest part of the summer. Grama grass may brown in extended periods of hot, dry weather but quickly greens again after a few good rains. Folks in areas receiving less than 8 inches of precipitation annually may need to water every 2 to 3 weeks during the summer to keep grass alive and healthy.

- ***Fertilizing*:** Apply a single application of Gro-Power 5-3-1 in the early fall.

- ***Mowing:*** This can help thicken a new lawn. Two months after planting cut the grass to a height of 3 to 4 inches. An established lawn can be cut 1 or 2 times to a height of 3 to 4 inches over the course of the summer if you want a more manicured look. Mowing can also be avoided, if you wish.

Legacy® Buffalo Grass

This particular native grass is well suited to poor soils and gives a nice cooling texture in any landscape. It's especially adaptive to clay soils.

Typically Plugs Are Used

- If your plugs are greenhouse-fresh, they need to be hardened off before planting. Place the flats outside in a morning sun/afternoon shady area for 7 to 10 days, leaving them outside at night.
  If a frost is expected, simply cover the trays with plastic or a blanket for the night.

- Water as needed. Hardening off in cold weather will cause the plugs to change color from green to brown or greenish-brown. This is OK; the grass will re-green as the weather warms. Once hardened, the plugs are ready to plant.

Preparing Bare Ground

- Be sure to remove all weeds from the site.

- Proper soil preparation is essential when planting Legacy® Buffalo grass plugs. The following soil improvements should be used for best results.

- ***Superphosphate 0-18-0:*** Use 2 lbs. per 100-sq. feet of bed area. Bone meal or rock phosphate may be used instead, at rates recommended on the package. This is an essential nutrient for strong root systems and is usually available locally.

- ***Planter II Trace Mineral Fertilizer:*** Use 2 lbs. per 100-sq. feet of bed area. This organic soil builder boosts essential trace mineral levels and increases microbial activity in the soil to improve nutrient availability.

- ***Gro-Power 5-3-1:*** Use at the rate of 10 to 15 lbs. per 100-sq. feet of bed area. Gro-Power is a humus (compost) based fertilizer that adds much needed nutrients and beneficial microbes to the soil.

- ***Compost:*** If available, add a well-made, thoroughly rotted compost to the soil up to a rate of 1/2 cubic yard per 1000-sq. feet. When adding compost, the amount of Gro-Power 5-3-1 used can be decreased by half or eliminated altogether. Rototill all the ingredients into the soil to a depth of 4 inches. Water thoroughly 3 or 4 days before planting to settle the soil and dissolve the Gro-Power granules.

Planting on a Slope

- Cover the soil with an erosion-preventing material, such as burlap or aspen excelsior matting that should be pegged into place. The plugs are planted through the material and the grass runners will root through it as they spread.

Planting Grass Plugs in Existing Turf

- Before planting, water plugs lightly.

- Make a single shallow 1/8 inch deep cut through the surface roots on each side and the bottom of each plug.

- Drill hole slightly less than the depth of the plug.

- Plant the plugs 6 to 15 inches apart in a grid pattern. The closer they are planted, the more quickly the new grass will fill in the area.

- Layer a 1 inch blanket of mulching material to keep it moist between watering and discourage weed growth.

- When mixing Blue Grama and Buffalo grass, plant the Grama plugs randomly amongst the Buffalo Grass plugs.

Watering Newly-Planted Plugs

- Water planted plugs thoroughly. The frequency will depend on how quickly the soil dries. Water just enough to keep the soil damp (not muddy) to a depth of about 2 inches. Initially, this may require a daily regime (morning or late afternoon) for the first week or so. As the plugs begin to root, watering can be reduced gradually to every 2 to 3 days, then to 1 to 2 times a week. (Plugs that are rooting will be noticeably greener and larger than those that haven't caught hold).

  After the first month, if it's not too hot and dry, the plugs should require only weekly waterings to keep them growing vigorously. *This is only a suggested watering schedule.* You should check the depth of soil moisture in several spots around the new lawn area before watering. It is better to let the plugs go a little dry for a day or two than overwater them.

- ***Gardening Tip:*** Water sloped areas with a fine spray, just enough so the water doesn't run off. Wait until the surface water is absorbed, then re-water 2 to 3 more times until the soil is wet to a depth of 2 inches.

Maintenance

- ***Watering:*** Once established, Legacy® Buffalo Grass is very drought tolerant, but may need extra water during the hottest part of the summer to keep the plugs green and actively growing. When depending on natural rainfall, grass may brown in hot, dry weather but will green-up quickly after a few good rains. If your area receives less than 8 inches of precipitation annually, you may need to water every 2 to 3 weeks during extended hot, dry periods.

- ***Fertilizing***: Legacy® Buffalo Grass has a higher need for fertilizer than Blue Grama. Apply two applications of Gro-Power 5-3-1 in early fall and late spring at the rate 1/2 lb. per 1000-sq. feet of grass. When establishing your Legacy® lawn, it is desirable in the first growing season to provide extra nitrogen. A monthly top-dressing with nitrogen-rich Soil Mender® Compost or Gro-Power 5-3-1 will improve the speed and density of fill-in.

- ***Mowing:*** This can help thicken a new lawn. Two months after planting, cut the grass to a height of 3 to 4 inches. An established lawn can be cut 1 to 2 times during the summer if you want a more manicured look. Mowing is not a necessity and can be avoided altogether, if you wish.

Planting Perennials

Perennials are plants that return to grow again in the spring after going dormant in the winter. If cared for, perennials will last many years. They don't grow as rapidly as annual flowers and most require 2 to 3 growing seasons to reach maturity. Planting in early spring or fall gives the benefit of cooler weather and will require less watering than those planted in the summer months.

Soil Preparation

- Water the area where you will be planting.
- Plants grow best in soil that has been loosened and enriched by digging in compost and other amendments.
- If planting individual plants outside of a prepared planting bed, dig a hole at least 12 inches deep by 12 inches wide for each plant.

- ***Gardening Tip:*** Be careful not to over-amend the soil when planting Penstemons and other native plants that prefer poor and humus deficient soil. Instead, loosen the soil and add Planters II and Superphosphate at the recommended rates.

Preparing the Site

Dig a hole twice as wide and a little deeper than the pot the plant is already in. Have extra compost and topsoil handy in case the hole is actually larger than the plant's root ball. Mix soil amendments into the soil at the base of the hole.

Preparing the Plant

- Remove plants from their pots by turning them upside down and lifting the pots off the root ball. If it won't slip off easily, gently squeeze the sides of the pot to loosen the plant. Pulling plants by their stems and leaves can cause damage.
- Prepare the roots (called scoring, scratching or roughing out) by cutting into the root ball about 1 inch deep. Make vertical cuts, top

to bottom, on each side of the root ball, and several cuts across the base. You are not hurting the plants by cutting into the roots.

Then with your fingers, loosen the roots. This ensures that the roots grow vigorously into the surrounding soil and do not continue growing in the shape of the pot in which the plant was grown.

Planting

- Install plants immediately after bringing them home. Early spring and fall planting gives the benefit of cooler weather and will require less watering than those planted in the summer months.
- Place the plant in the planting hole, being careful to set the top of the root ball even with the surrounding ground.
- In dry, hard-to-water areas, a shallow 1/2 inch depression below grade of the surrounding soil will help hold water.
- Don't pack the soil too tightly; and to prevent compacting, never water the soil as you are planting.

Mulching

Blanket a 2 inch layer of mulch around the plant to keep the roots cool and the soil moist and protected from the wind and sun. Crushed 3/8 inch gravel is a good choice for some xeric plants.

Watering a New Planting

- Perennials tend to be deep rooted, so let water soak deeply by creating a well around the plant to hold water so it can seep in.
- Use a root stimulator to prevent transplant shock.
- Don't let the soil around newly-planted perennials dry out completely during the first month after planting. However, properly planted and mulched perennials should not require daily watering. Initially, watering every two or three days may be necessary. After the first 4 to 6 weeks, gradually cut back watering frequency as the plants get established.

Year-round Schedule for Watering Perennials

Following are suggested frequencies for gardens with properly enriched soil that has been correctly mulched to protect it from the drying effects of the sun and wind.

- ***January – February:*** Water once every 2 to 4 weeks in full sun areas or other sites (like slopes) that dry out quickly. Winter watering is particularly important for plants going through their first winter.

- ***March:*** Continue watering once every 2 to 4 weeks as above.

- ***April:*** Increase watering frequency as soil begins to thaw and dry, once every 1 to 2 weeks.

- ***May:*** Increase watering frequency as days get warmer, once every 5 to 10 days, as needed.

- ***June – August:*** Soak deeply every 5 to 7 days for established plants, and more frequently for newly-transplanted plants. Remember, it is better to water deeply to encourage root growth.

- ***September:*** Regular deep watering every 5 to 7 days is important during this generally hot, dry month. Pay careful attention to new transplants.

- ***October:*** Reduce watering frequency to once every 1 to 2 weeks as the weather cools.

- ***November – December:*** Reduce watering frequency to once every 1 to 3 weeks.

Planting Ornamental Grasses

Much valued in the landscape for their elegant foliage and showy plumes, the ornamental grasses encompass a wide variety of textures, forms, sizes and colors. They capture light and movement in ways other perennial plants cannot. With changes in color and character throughout the year, they bring multi-season interest to any garden.

Soil and Planting Preparation

Planting requirements and care for ornamental grasses are the same as for perennials.

Cultural Requirements

As a broad classification of plants, ornamental grasses vary widely in their sun, water, and soil requirements. Generally, they require full sun and can tolerate part shade. Some grasses are more tolerant of drought than others, but all require regular water their first season while they are getting established. Most grasses grow well in a variety of soil types, as long as they are well drained.

Maintenance

- Cut back hard, to a height of 3 to 6 inches, in the spring to give new foliage room to grow.
- Mature grass plants will re-grow to their full height in a single season.
- The middle of the clump should be scratched out to remove dead stems and leaf blades.
- Leave grasses standing over the winter for ornamental interest.
- Divide big clumps every three or four years to re-invigorate the plant.
- The soil in which grasses are planted should be enriched with compost. Grasses also love some extra nitrogen or Soil Mender® Blend, and should be top-dressed with Yum Yum® Mix fertilizer in mid-to-late fall.

Planting a Creeping Thyme Lawn

Thyme makes a nice low-maintenance groundcover, and is good for planting at the base of taller perennials. Or it can be used for planting large patches as a thyme "lawn."

Cultural Requirements

Thyme is easily grown in light, well-drained soil that is kept fairly dry. Although it is mostly drought-tolerant once established, it will need thorough, infrequent watering. Check the soil moisture before watering, and if needed, water to a depth of 4 to 6 inches, then allow the soil to dry. This should be done once every 7 to 14 days, depending on the soil and weather conditions. Remember, thyme will need some winter moisture as well.

- Thyme grows best in full sun, though some varieties (Albus and Coccineum) will take part shade.
- Thyme is usually deer and rabbit resistant.
- Varieties of creeping thyme are generally cold-hardy to zone 4.

Replacing an Existing Turf Grass Lawn

When replacing an existing lawn with thyme, it is extremely important to completely kill the grass. Grass species like Kentucky Blue or Bermuda grass that spread by runners can re-establish themselves and eventually overgrow the thyme.

Use Round Up or Finale (non-selective herbicides that kill annual and perennial weeds, including the roots, without contaminating the soil) to kill the existing grass, roots and all, before preparing the soil.

Follow directions on the labels carefully. Give yourself plenty of time, Round Up requires 10 to 14 days to thoroughly kill lawn grass. Once the grass is dead, it can be removed by using a sod cutter to strip the sod off or by rototilling the dead grass and roots into the soil.

Proper Soil Preparation

When planting new thyme areas, the following soil improvers should be added.

- ***Hi Yield Superphosphate 0-18-0:*** Use 2 lbs. per 100-sq. feet of bed area. Bonemeal or rock phosphate may be used instead, at rates recommended on the package. This is an essential nutrient for plentiful flowers and strong root systems.

- ***Planters II Trace Mineral Fertilizer:*** Use 2 lbs. per 100-sq. feet of bed area. This natural soil builder boosts essential trace mineral levels and increases microbial activity in the soil to improve nutrient availability.

- ***Yum Yum® Mix:*** Use at the rate of 4 lbs. per 100-sq. ft of bed area.

- ***Compost:*** When available, add a well-made, thoroughly rotted compost to the soil at the rate of one cubic yard per 100-sq. feet of bed area. The compost and soil minerals should be tilled to a depth of only 6 inches, and not 12 inches or more as recommended for perennial flower beds.

Planting the New Thyme Lawn

Plants can be spaced from 6 to 12 inches apart for quick coverage. To remove thyme plants from their pots, squeeze the undersides of the pot and push up on the bottom of the root ball with your thumb.

When planting in hot, dry areas, work some water-holding crystals into the individual planting holes to improve transplanting success. With proper soil preparation, you can expect the plants to fill in and cover the area in approximately 6 to 8 months, depending on weather conditions.

Different thyme varieties can be mixed to vary the blooming times and texture of the thyme lawn. When doing this, be sure to plant the different varieties in groupings of 5 or more plants, for best results. Reiter, Pink Chintz, Coccineus, and Albus are the preferred varieties for use in a thyme lawn.

Watering a Thyme Lawn

After the new plants are in the ground, water thoroughly and use a root stimulator to reduce transplant shock and promote growth. Depending on how hot the weather, plants will need a good soaking approximately 1 to 3 times a week, for the first 2 to 3 weeks. Once the plants begin to root, watering frequency can be cut back to a good soaking once every 7 to 10 days.

Once established, thyme is drought tolerant and it's better to keep it on the dry side. When watering, be sure to soak the soil to a depth of 4 to 6 inches when the plants are dry.

Yellowing foliage can be a sign of overwatering.

Mowing a Thyme Lawn

To keep a thyme lawn looking tidy after blooming, it can be mowed to remove the faded flowers and to help the stems fill in any bare spots.

Fertilizing and Maintenance

Fall is the optimum time to apply fertilizer. A single application of Yum Yum® Mix applied at the rate of 2 lbs per 100-sq. feet in mid-to-late fall (late Sept.- early Nov.) will keep the lawn looking good. Be sure to water in the Yum Yum® Mix with a good deep soaking right after you spread it.

If the plants have flowered heavily and the foliage is a little thin after it's mowed, a light shot of Earth Juice or a water-soluble type like Miracle Gro is recommended to help speed fill-in.

A light raking in the spring can be helpful in removing dead stems and foliage after a harsh winter. Then top-dress with a thin 1 inch of finely textured compost or well rotted manure to help the plants spread to fill in bare spots and reinvigorate the whole lawn for the coming of summer.

Walking On a Thyme Lawn

Some thyme plants are very tolerant of foot traffic. However, paving stones or flagstone pieces are recommended for pathways and other areas that will receive heavy foot traffic.

Planting Guide for Trees and Shrubs

This guide is good for conifers and deciduous trees, as well as all shrubs.

Things To Do Before Planting:

- Call utility company and have utility lines marked before digging.
- Make sure area is accessible for driver to deliver large trees.
- Mark irrigation lines near planting area.
- Have water available to planting area.

Planting

- Dig hole twice as wide as container and the same depth.
- If ball & burlap (b&b), dig just deep enough to keep top of root ball at ground level.

Prepare Backfill

- Mix soil from hole with Soil Mender® Blend at the rate of 1 part to 2 parts soil.
- Add Superphosphate; and if needed, add soil sulfur.
- Recommendation for superior growth: Add a complete fertilizer such as Yum Yum® Mix with Planters II.

Prepare Root Ball

- Do not break root ball.
- In container cut 5 or 6 slits around the edges and bottom of the root ball to encourage new growth.
- If the ball & burlap comes with wire basket, leave entire wire basket on until placed in hole.

Set Tree in Hole

- Center tree in the hole to proper planting depth.
- On ball & burlap with wire, cut away wire on sides as deep as possible.
- On ball & burlap only, untie burlap and pull burlap away from top of root ball — do not remove the burlap! Doing so may cause damage to the root ball.
- Add amended backfill to hole.
- Tamp down new soil to remove air pockets.
- Create water basin around plant.

Watering New Plantings

- Add water slowly to saturate root ball.
- Add Root Stimulator Combo (Superthrive and liquid Seaweed Concentrates).
- Repeat application of Root Stimulator in 2 weeks.

Maintenance Watering

- Water regularly for the first two growing seasons to establish root systems.
- Water at least every 3 to 4 weeks in winter.
- Use 2 to 4 inches Soil Mender® Mulch year-round on top of the soil, inside the water basin.
- Fertilize in the fall as needed.

Staking

- Most trees *do not need* to be staked at planting time. Staking is only needed if a tree is in danger of blowing over. Tall aspen and large, densely needled conifers are the most likely to need it.

- Drive two stakes into the ground so the main trunk of the tree is in line with the stakes.
- Use special tree ties or cut sections of hose held with wire to the stakes, to hold the tree trunk. Secure ties to the stakes.
    - When using hose, don't run wire (or twine) through the hose. Loop the hose around the trunk and secure the ends of the hose.
- Remove stakes and supporting tie after a year. If left in place, staked trees are actually weakened by not standing on their own.

Curly Leaf Mountain Mahogany

Special Instructions for Planting Live Christmas Trees

Shelter

Keep your live tree outside in a wind-free location. Water the tree every 7 days, and water it thoroughly the day before you take it inside the house.

Preparing to Plant

Dig the planting hole before the ground freezes. Pre-mix the backfill soil with amendments then mound it and cover it with plastic. Be sure to cover the hole with boards, or fill it with mulch, to prevent an accidental fall.

Inside the House

Do not keep your tree inside the house for more than 7 days. This is very important. A tree responds to low winter temperatures by going dormant and ceasing new growth. If left inside a warm house for more than a week, the tree thinks spring has arrived and will break dormancy before winter is over. It can easily freeze during the remaining cold weather.

If possible, set your tree near a cool window. Keep it away from fireplaces or other heat sources, and use only small, low-voltage Christmas lights on the tree. Water enough to keep the root ball slightly moist.

After the Holidays

If possible, leave your tree in a wind-sheltered "transition zone" for 2 weeks before planting it. This zone can be an unheated garage or a protected corner of the yard. Continue to water the tree weekly.

After planting, water the tree weekly for 2 weeks, then once every 2 weeks until spring.

If you don't plan to plant the tree until spring, keep it in a wind-sheltered location. You can sink the container in the ground and mulch it up to the rim to prevent freezing. Water every 2 weeks.

Planting Wildflower Seeds

There is nothing like a full meadow of wildflowers that you've planted yourself. In our premixed bulk Grass and Wildflower Re-Veg Seed Mix the grasses are Wheat, Rye, Brome, Fescue. The flowers are Flax, Black-eye Susan, Coreopsis, Penstemon, Wallflower, Bachelor Button, Firewheel, Poppy and Yarrow.

When to Plant

Any time in late fall or early spring (when the soil is not frozen) is fine. However, if you're planting in the fall, make it late enough so the seeds do not germinate until spring.

Preparing the Soil

Clear area of all weeds. Then breakup the top 2 to 3 inches of the soil and amend with organic matter, such as Soil Mender® Blend, for good root and water penetration.

Smooth the Surface

Lightly rake the area to even it out.

Scatter Seeds

Use 1 oz. of seed for every 125-sq. feet or 1 lb. for every 2000-sq. feet Blend seeds with sand at a ratio of 1 bucket of sand per 1 lb. of seed. Sow either by hand or with a seed spreader. It helps to sow the first pass in one direction; then crosshatch the second pass to ensure better coverage. But don't over-seed and make a crowded situation. Reseeding during the summer months may be necessary if germination was erratic and there are bare spots.

Rake and Set Seeds

Using your hands, the back of a bamboo or fan rake, lightly brush the seed under the soil. It's important for seed and soil to make contact for germination.

Mulch

Lay a thin dusting of rich organic soil (Soil Mender® Top Soil) or compost (Soil Mender® Blend) to provide shade for new seedlings and help retain moisture. This also keeps the birds from eating the seeds.

Water

When April rolls around, it will be necessary to water if winter/spring precipitation has been lacking. Water thoroughly so the soil is damp to a depth of 3 to 4 inches. Once watering has begun, keep the soil surface damp for the first 3 weeks.

After the emergence of seedlings, be careful not to overwater; this will kill the seedlings. For the second 3 weeks, reduce the frequency until you are watering deeply to a depth of 6 inches once a week. By now the wildflowers should be pretty well established and can make it on their own.

Weeding

Weeds can quickly take over a new planting. However, be careful not to pull the wildflower seedlings.

- ***Gardening Tip:*** Germinate a pot full of the seed mix you've planted so you can identify the seedlings of the wildflowers. Once the seedlings are large enough not to crush by walking on them, hand weed every few weeks.

Planting Winter-hardy Cacti and Succulents

Today, many species of cacti and succulents are close to extinction in their native habitats due to irresponsible collectors. So please don't collect cacti from the wild unless it's to rescue plants from construction sites.

Choosing the Site

- Select a bedding site with full sun and preferably next to a south- or west-facing wall; these add extra warmth during the winter.

- Plant on a slope or raised area, not in a low spot that collects water.

Preparing the Soil

- All species of hardy cacti and succulents require fast-draining soil.

- In heavy clay soil, replace half the soil from the hole with coarse sand mixed with the remaining soil. This ensures adequate drainage.

- Do not add compost; add only a small handful of Planters II.

- In an outdoor pot or planter, use a potting mix of 3 parts potting soil; 1 part coarse sand; 1 part medium or coarse perlite or similar material.

Planting

- Cacti and agaves should be transplanted bare-root. Let the soil in the pot dry out for a few days. Then remove the pot and gently loosen the soil so it falls away from the roots. Trim off any broken roots. Then evenly spread out the roots; set plant in a shallow hole and sprinkle soil into the hole until full. The base of the plant should rest on top of the soil.

- Mulch with a 1/2 to 1 inch thick layer of pea-sized gravel around the base of the plant to protect it from contact with soggy soil over the winter months.

- Succulents (Aloinopsis, Titanopsis, Ruschia, Delosperma, Sedums and others) need not be transplanted bare-root; instead, the root ball should be scored and roughed out like other perennials.

Watering

- Bare-root cacti *must not* be watered right away, but should sit dry for a day or two to allow the roots to callus over any broken or damaged areas. Succulents can be watered right away. All plants should be watered thoroughly after planting with a mixture of Saltwater Farms liquid seaweed and Superthrive to stimulate strong new root growth. Water again with this mixture 2 weeks later.

- Initially, water outdoor beds with new plants once every 5 to 7 days for the first month or so after transplanting. Cacti and succulents enjoy regular watering during the heat of the summer and will grow vigorously. After the first year, most cacti species need a good soaking once every 2 to 4 weeks during the spring and summer, if there has been no rain.

- Potted plants require more frequent, regular watering, especially if the weather is hot and dry.

- To prepare outdoor cacti and succulents for winter, begin withholding water in the fall so the plants can begin to dehydrate and shrivel. Plump, well-watered plants will be damaged when temperatures plunge.

Winter Protection

- Many cacti and succulents are quite cold-hardy if kept dry in the cold winter and spring months. In areas that receive a lot of winter and spring moisture (especially rain), it's strongly recommended that plants be protected from cold and wet conditions.

- Problems will occur if plants sit in wet soil all winter or even under melting snow for extended periods.

- Set potted plants under an overhang on the south or west side of the house, or place in a well-ventilated cold frame. Water pots and other containers lightly a few times over the winter during warm spells.

- ***Gardening Tip:*** A temporary cold frame can be constructed using wire hoops covered with a clear plastic sheet to cover the entire bed. Or individual plants can be covered with plastic gallon milk jugs with the bottoms cut out to keep the ground around the plants dry. Leave the top off the jug so heat build-up isn't excessive.

Moisture Tolerance

- The most moisture-tolerant cacti species (best adapted for growing outdoors in areas where cacti are not native plants) include: Coryphantha vivipara, Echinocereus reichenbachii varieties, Echinocereus viridiflorus, Pediocactus simpsonii, Escobaria missouriensis, and various Opuntia (pad cacti) species.

- South African succulents are very sensitive to wet soil in freezing weather. They particularly dislike being covered with snow for extended periods and will rot. When kept dry, these plants have excellent cold tolerance.

- Plant them wedged between rocks, in sloped south and west-facing beds where snow melts quickly and the soil is very well drained. In cold, wet winter climates it is best to grow these plants in a container so they can be moved to protected areas. Also, be on the lookout for hungry rabbits; they will occasionally nibble on these succulents.

Fertilizing

Cacti and succulents are very modest in their fertilizer requirements. If planted in the ground, fertilize in spring with Saltwater Farms liquid seaweed and Yum Yum® Mix. If planted in pots, feed monthly with the same mixture as above, beginning in late spring and continuing through late summer.

Growing Xeric Plants in High Rainfall Regions

This also includes cacti and succulents. Growing plants that are adapted to drier conditions in wetter areas of the country is both challenging and satisfying. The key to success is to provide optimum growing conditions. If you grow lavender, then you are familiar with the growing conditions that other xeric plants need.

Soil

Xeric plants like a lean, low fertile soil. They do well in poor soils that aren't amended. If you have rich planting beds, you'll have to augment with sand and gravel. Add lime when soil is acidic.

Drainage

This is the most important aspect to consider. Roots of xeric plants can't stand around in wet, soggy soil. They need water to drain quickly. The more rainfall your area receives, the sandier the soil needs to be.

- Test the soil/sand mix. Fill a hole with water to saturate it. Refill 10 to 15 minutes later. If water doesn't drain from the hole in 30 seconds, add more sand.

To create good soil drainage, follow this recipe:

- Loosen soil with coarse sand (at least 1 part sand to 1 part native soil).
- Clay soils need to be amended with 3 to 4 parts coarse sand to 1 part clay; pile into a berm (small mound). Or remove the clay to a depth of 8 inches and re-fill with clay/sand mixed with Planters II. Planters II is the only soil amendment needed. Don't add peat moss, compost or other humus containing amendments.

- ***Gardening Tip:*** Keep soil bare and unmulched, or mulch with crushed gravel. Gravel protects the crown from excessive winter moisture, which is the enemy of xeric plants. Gravel also encourages plants to re-seed.

Sunlight

Xeric plants love full sun. They also thrive in very warm places, so if possible place these plants next to south-facing walls, near boulders and along brick or flagstone walkways. They'll pick up the radiated heat.

Site Placement

Situate these plants high. On top of a berm (mound) or in a raised bed is ideal. This allows water to drain easily. Plant with other xeric plants; don't mix plants that like more water with those that don't.

Fertilizing

If your soil is rich, you won't need to fertilize. If you feel a need to fertilize, an annual application of Yum Yum® Mix in mid-fall is recommended. Don't use compost as this will increase the soil's humus content over time. Xeric plants don't like rich soils.

Watering

If you live in an area with more than 25 inches of water a year, additional watering won't be necessary after the plants have established themselves the first growing season.

African Blue Basil

Notes for Planting and Caring for Plants

Around Your Garden

THIS SECTION includes information about gardening year around. Also, due to the increased danger of fire in so many parts of the country because of drought, a few tips about firescaping are included.

Know Your Zone

Paying attention to the USDA Plant Hardiness Zones lets you choose plants that are appropriate to your locale. Hardiness zones refer to minimal air temperatures reached during the winter. If temperatures are too cold, some plants won't survive. A plant designated for zone 7 can't be expected to live through a winter in zone 4; the plant isn't able to produce the chemical processes needed to ensure winter hardiness.

However, if a freak cold snap hits in a warm climate, the low temperatures may cause cellular damage, but may not be fatal to plants if the low temperatures don't persist.

If you're unsure of how cold it's gotten in the past, contact your local County Extension Service for information. Below is a hardiness zone reference with degrees in Fahrenheit.

Zone	Temperature	Notes
1	Below -50	
2	-50 to -40	
3	-40 to -30	
4	-30 to -20	soil freezes solidly and deeply from late fall through early spring
5	-20 to -10	soil freezes more deeply for the winter
6	-10 to 0	soil will only freeze down 1 to 3 inches for a few weeks at a time
7	0 to 10	soil rarely freezes more than just below the surface
8	10 to 20	
9	20 to 30	
10	30 to 40	
11	Above 40	

Fall is for Planting

We recommend planting many species of perennials, trees and shrubs in the fall from early September through late October or until the ground freezes.

Why Plant in the Fall?

- 80% of a plant's root growth occurs in the late summer and fall months.
- The air is cooler, so plants don't need as much water.
- There is less transplant shock because of the cooler air.
- After the tops of plants have stopped active growth, all the energy of a plant returns to the roots. It's during the fall that roots get a chance to absorb all of the nutrients without needing to distribute them to leaves and blooms.
- Roots get an extra few months to absorb all the nutrients they need to make it through the winter. With a more established root system, fall-planted trees, shrubs, and perennials are better able to handle the harsh, drying winds of spring and the withering heat of summer.

Fall Planting is a Waterwise Strategy

- As plants begin to go dormant in the fall, they use less water.
- The soil is cooler in the fall, so it stores moisture better.
- Watering is easier because you have to water less frequently. Fall is less windy than spring. Wind dries out the soil more quickly and dehydrates plants.
- Water regularly through the fall, decreasing frequency as daytime temperatures cool in late October and November. From 2 to 4 weeks after planting, a plant should be rooted, especially if Superthrive root stimulator is used at the time of planting. Once a plant is established, watering frequency can be reduced.

- After the soil begins to freeze, soak plants once every 2 to 3 weeks through the winter months, except when it is very cold and the ground is frozen solid.

Fertilizing in the Fall

- Fall is the time to apply a fertilizer that promotes root development, but not new top growth.

- Choose a fertilizer that is low in nitrogen but high in phosphorous (see NPK in glossary).

- For newly planted perennials, use a root stimulator to help plants avoid transplant shock and encourage the development of a strong root system.

- Top-dress planting beds with Planters II, an organic trace mineral fertilizer.

- Add Greensand to break clay soils.

- Add soil sulfur for all acid-loving plants.

Correcting Iron Deficiencies

If plants have yellow (chlorotic) foliage, they may be suffering from an iron deficiency. Fall is the best time to increase the amount of available iron with a top-dressing of Greensand, a rich source of iron and potassium. Also add soil sulfur. This lowers the soil's alkalinity and makes iron more readily available.

- ***Gardening Tip:*** For chronically iron-deficient plants, dig a ring of 1-foot deep holes around the drip line of the tree and add a mixture of Greensand and soil sulfur, along with generous amounts of compost into the holes.

Exceptions to Fall Planting

Plants that are borderline in their cold hardiness should not be planted in the fall.

- In zone 3 to 5, trees and shrubs that would best be planted in spring include: Rose of Sharon, Crape Myrtle, Photinia, Nandina, Mahonia, hybrid roses, Mimosa, Japanese Pagoda, Ginko biloba birches, Red Bud and broadleaf evergreens.

- Perennials to avoid planting in the fall are: Agastache, Salvia greggii, Verbenas, Zauschineria cacti and South African succulents.

Fall Clean-up and Winterizing Your Garden

Besides planting, fertilizing and mulching in the fall, there's cleaning up. It's always amazing this time of year there's so much to do to get a garden ready for winter.

- After a hard freeze, plants typically drop their leaves. The leaves that cover lawn areas and xeric plants should be cleaned up so they don't shade-out the lawn or keep the crowns of xeric plants too wet over the winter. Add leaves to your compost pile or put them near trees and shrubs as mulch.

- Leave the stems of perennials because most plants over-winter more successfully when the stems are left intact. Stems provide energy reserves for the root system and catch blowing snow that helps insulate a plant.

- Bring houseplants inside that have summered in the garden, and up-pot the ones that have outgrown their containers.

- Drain garden pools and birdbaths, unless you have a water heater that keeps the water from freezing. Remove water plants; seal root ball in a garbage bag and store in a cool, non-freezing place.

- Prune dead branches on trees beginning mid-November. Torn wounds and dead wood are places where insects like to hibernate and lay eggs. Climbing roses with very long canes may be trimmed back so they don't whip in the wind. Tree branches that have narrow branch angles may be prone to breakage in heavy snows and should be pruned.

- Spray Deer-Off repellant on the twigs (and foliage, if the plant is evergreen) to discourage browsing.
- Fall is an active time for gophers. Watch for fresh soil mounds to determine if they are eating plant roots in your landscape. Mole and Gopher-Med is an excellent repellant that is sprayed onto the top of the soil and watered in.
- For new broadleaf evergreens install simple windbreaks or wrap in burlap to protect from winter weather.
- After a heavy frost, remove all blackened annual plants.
- Clean garden tools so they will be ready for next season; get them sharpened too. There is nothing like starting out with clean, sharp tools in the spring.
- Properly store seeds, fertilizers and garden chemicals where they won't freeze.
- Drain and store garden hoses. However, be prepared to winter-water first year transplants.
- And if you have one of those wonderful old wooden gates in your garden, give the hinges a shot of oil to reduce winter rusting.

Winter Watering

The most common reason for plants to die off during the winter is not the cold, but the lack of water. This is particularly true in the arid western US. Cold-hardy plants can make it through brutally cold weather, but they can't go without water. But just because you can't stick your finger in the soil doesn't mean water can't seep in. And seep in it must.

Why Water Now?

Dormant plants need moisture to maintain their physiological and biochemical changes that makes them tolerant of cold. Water helps produce high concentrations of dissolved sugars, amino acids and other soluble organic molecules. Moisture-laden plant cells help the elasticity of protoplasm, which in turn makes plants resilient during freezing temperatures.

It's easy to assume plants aren't doing anything during the winter, but a lot we can't see is taking place at the root level.

Physics at work in the Winter Garden

- Even if the soil is frozen, it will absorb water. When water meets the ice, the crystals gives off a crackling sound as it soaks in. Dissolving ice helps aerate the ground.

- Make sure the soil has good drainage to prevent heaving. Heaving (or lifting of the soil) occurs during the thaw-and-freeze phases. Heaving can fracture taproots and lateral roots, sometimes resulting in elevated crowns which quickly dry out a plant. A thick application of mulch in the fall will help keep the soil frozen and prevent heaving.

Winter Watering Guidelines

Continue watering into the fall, but at less frequent intervals than during the summer. This means watering thoroughly once every 2-4 weeks and only if the air temperature is above freezing. Then depending on the amount of precipitation, this watering schedule can be kept up throughout the winter.

- Water thoroughly once every 2 to 4 weeks. Sandy soils need more frequent watering than heavy loam and clay soils.

- Try to water early in the day so the moisture has time to soak into the ground before freezing again at night.

- In colder regions where the soil freezes deeply, there is no need to water during the coldest intervals of winter. Just make sure the soil is damp going into late fall.

- In milder winter climates, and particularly in the Southwestern US where it's sunny and the soil doesn't freeze deep down, winter watering is essential.

- Don't over water. If winter snows/rains have left the soil moist, no supplemental watering is needed. Keep in mind that sodden roots are unable to gather water, which in turn starves the plant for oxygen and other nutrients.

- Be sure to water during February/March thaws.

- Drain hoses so you can water again.

- When watering woody plants that have been planted in the last two seasons, the most important area to water is about a foot from the drip line. It's this area that needs to be kept moist because the root hairs spread horizontally. Doing this also prevents water from building up near the trunk of the plant where it can freeze and damage the bark.

•`***Gardening tip:*** A rule of thumb is when a half-inch of top soil on a winter garden is loose, that's a good time to water. Even if soil is frozen, it will absorb water. The dissolving ice helps aerate the ground.

A Discussion of Pruning

Winter and early spring months are a good time to check plants for pruning needs. Without leaves, growth problems are more evident: branches too numerous, too close, a forked leader or going in the wrong direction. Plants in dormancy are also not expending energy so their energy levels for recovery are highest.

Pruning Times

Deciduous Plants

Fruit trees, flowering trees, shade trees, summer flowering shrubs and vines all favor a late winter, early spring pruning.

- ***Gardening Tip:*** The worst time to prune a deciduous plant is when everyone wants to prune it—in the spring when it's putting on new shoots. Pruning a plant during a growth period can shock it enough to cause dieback.

Conifers

Shape evergreens such as spruce and pine during their short growth period in late spring, early summer. To contain a conifer's growth, cut developing candles back by half before the needles grow to their full length. Doing this yearly will encourage a thicker tree. It is much better to contain an evergreen's size by yearly shoot pruning than to wait until the tree is too large for the space.

- ***Gardening Tip:*** Removing horizontal growth channels plants' energy to the remaining upright branches, allowing it to grow into a taller specimen.

Helpful Tree Pruning Hints

For trees and shrubs, it is best to remove no more than a third of the plant's growth at a time.

- Pruning dead or diseased wood can and should be done any time of year. If there is any chance of disease, tools should be sterilized with a 10% solution of bleach after pruning each plant. If branches are infected by Fireblight, tools should be sterilized after every cut.

- Use tools that are sharpened, oiled, and the right size for the job. A sharp tool that is large enough for the job will make a clean cut. Jagged pruning cuts attract insects and disease. Be sure to prune trees at the branch collar (the small swollen ring at the point of attachment to a larger branch), not flush to the trunk.

- Pruning deciduous trees in fall at the time of leaf drop should also be avoided.

- Pruning paint is not recommended to seal a fresh cut. The plant sap is effective enough.

- If a limb of an evergreen is cut back to a point where there are no longer any green needles, it is unlikely that the plant will re-produce growth from the cut.

Pruning Hints for Shrubs

- Flowers on shrubs such as Lilacs, Forsythia, and Climbing Roses bloom on last year's growth. Any of this growth removed in early spring will consequently result in a loss of flowers. For maximum flower show, prune spring flowering shrubs after they bloom.

- Summer blooming shrubs such as *Potentilla,* Butterfly Bush *(Buddleia),* Russian Sage *(Perovskia)* and cold hardy shrub Hibiscus should be pruned in mid-spring, giving the plant plenty of time to re-grow summer flowering branches.

- In areas with zone 5 or colder winters, roses are best pruned in mid-April to early May. Pruning too early stimulates new shoot growth, which can be damaged by late frost. Roses should be pruned using the less than one-third removal rule. Cold, short season climates limit time for re-growth; removing too much wood will stunt the roses so prune judiciously. Applying glue to newly pruned tips will keep the cane borer from damaging the rose cane.

- Avoid pruning shrubs into the shape of globes and squares. It destroys the natural beauty of shrubs and creates a work intensive situation that must be continued indefinitely. If the shrub is too large for its space, remove it.

- ***Gardening Tip:*** A carefully pruned plant looks better with an even distribution of branches, no short stubs or old canes at the base of the plant. Also cut root sprouts ("suckers") back to below ground level. Keeping a plant's growth cycle in mind, pruning will allow flowers and fruit to remain in abundance.

Getting Ready for Spring

Spring is a good time to clean up your perennial garden and prepare it for the coming growing season. Between mid-February and mid-April, you can cut back the previous year's dead growth, prune woody perennials and shrubs, and divide large or overgrown species. Spring is also a good time to fertilize.

Cutting back and pruning

- Any plant that sends up new flowering stems each year should be pruned back to the basal foliage (fresh leaves coming up from the base of the plant). These include Agastache, Asters, Beebalm, Catmint, Columbine, Coneflower, Coreopsis, Daisies, Helianthus, Hollyhock, Sages, Penstemon, Phlox, Sedum, and Yarrow. Care should be taken not to injure any new growth that has already emerged at the base of the plants.

- Depending on their mature size, ornamental grasses should be cut in the spring 3 to 12 inches from the ground. The clump should be raked lightly to remove dead stems.

- Woody perennials should be pruned heavily before they start to show signs of new growth. Lavender, bush Sages, Santolina, Rosemary, Erysimum (Wallflower) should be cut back one-third. Blue Mist Spirea (Caryopteris) and butterfly bushes can be cut back one-third every third growing season, depending on their size. Artemisia Powis Castle and Russian Sage can be cut 6 to 12 inches from the ground.

Dividing perennials

Perennials can be divided to keep them healthy and strong, or to propagate additional plants. Spring divisions should be done when the plants are dormant. Otherwise, wait until fall. Don't try to divide perennials with taproots. They should remain undisturbed.

- Divide a perennial when the center of the plant becomes open and weak, or when the plant's seasonal growth diminishes and it begins to produce fewer flowers. If a plant has overgrown its site, it may be better to move it to another location, rather than keep dividing it every few years as it continues to get bigger.

- To divide a perennial, dig up the entire plant, with as much of the root ball intact as possible. Separate the roots into two or more portions, depending on how quickly the plant grows. For example, if the plant grows very rapidly, like Beebalm, it can be divided into many small pieces. Replant the divisions in suitable sites, where soil has been prepared with compost or other amendments. Water deeply and mulch.

Fertilizing

If perennial beds were not fertilized in the winter, fertilizers can be added in the spring, at rates recommended on packages. Scratch quantities into the soil and water thoroughly.

- Use organic fertilizers to build your soil. Hardy garden perennials thrive in rich humus soils.

Mulching

- A fresh 2 inch layer of Soil Mender® Mulch will help in weed control and water retention.

Watering

- Water perennials once every 2 to 4 weeks during the first part of the year, and increase watering frequency to once every 1 to 2 weeks in April, as the soil begins to thaw and dry more quickly.

- Spring is a good time to rebuild or expand the size of earthen basins, or wells, around your plants. These are raised ridges of soil around the perimeter of your plants that hold water and keep it from running off.

- ***Gardening Tip:*** Cut the rim off nursery pots to arrange around plants for a more permanent well that won't wash away.

Firescaping for Peace of Mind

In recent years we've all become aware of fire danger. Around houses it's imperative to clear trees and other foliage for protection against wildfires. The basic guidelines for trimming around houses follow:

- **Zone One** is a 30-foot cleared area surrounding a house. Concrete or brick patios in this area are ideal as well as low ornamental shrubs. If trees are to be planted in this first zone, they need to be deciduous.

- **Zone Two** moves out another 70 feet and is good for orchards and gardens. Lower limbs of trees should be pruned to 15 feet off the ground.

- **Zone Three** is another 100 feet from the house. Crowns of trees should be separated by at least 10 feet. Prune branches to a height of 10 feet off the ground.

Further tips—

- Remove branches within 15 feet of chimneys and stovepipes.

- Create fuel breaks wherever possible with such items as pools, fountains and non-flammable fences. Lay rock, gravel, brick and paving in wide-open areas.

- On steep slopes, plantings need to be far apart to discourage fire from climbing up a hill.

- Reduce the fire fuel so it doesn't "ladder" up from the ground into trees, and use fire resistant plants wherever possible.

Notes Around Your Garden

Ground Cover
Veronica pectinata

Garden Glossary

Common Gardening Terms

Amendment: Any soil additive that increases its texture, drainage or humus.

Annual: Plants that complete their life cycles from germination to seed in a single growing season.

Beneficial insects: Bees, butterflies and moths that spread pollen among plants. These also include Ladybugs and Praying mantis that eat aphids and other harmful pests.

Cactus: Members of the cactaceae family; these desert-living plants are characterized by their thorns or spines and fleshy bodies adapted to storing water.

Cold snap: A sudden drop in temperature, often below freezing and usually during the spring. Cold snaps may cause damage to blooming and leafing plants.

Cold-hardy: This refers to perennial flowers, shrubs and trees that can survive cold and/or subfreezing temperatures and return for another growing season.

Chlorotic foliage: A condition in which the newer leaves of a plant turn yellow due to an iron deficiency or lack of some other mineral.

Companion plant: A plant that looks good when planted with another, either in terms of height, color or texture. It also refers to plants that have a beneficial effect on each other. For example, nasturtiums planted with squash will repel squash bugs.

Compost: A soil additive that is rich in humus of decomposed plant material.

Cool-season grasses: Grasses that start greening up and do their growing during the cooler months of spring and fall. They are the first grasses to turn green and will remain green late in the year. During the hotter months of summer, they go dormant and turn brown unless irrigated with lots of water. Cool-season grasses do best when planted in the spring.

Crown: The top of a plant at or just below the soil level that the stems grow from.

Cultivar: A cultivated variety from a plant that grows naturally in the wild.

Dead-heading: Cutting off the spent flower heads on plants after they have bloomed. However, if you want the plant to re-seed itself, leave the spent blooms intact on the plant.

Deep-watering: Soaking the ground around a plant so the roots are saturated.

Drip line: The area that lies under the outer branches of a tree, where the water drips from the leaves and needles.

Fast-draining: Soil that is loose and lets the water easily flow through it.

Fertilizer: Chemical or organic nutrients that are taken up by roots.

Greening-up: This refers to plants and shrubs that are beginning to put out new growth in the spring.

Harden-off: A process of gradually getting young plants used to being outside after germinating or being grown indoors or in a greenhouse. Too much sun, wind and low humidity can damage seedlings if they aren't acclimated to these outdoor conditions.

Herbaceous: This refers to plants that have soft tissues as opposed to woody stems. These plants also die back to the ground during cold months.

Leaf-out: This refers to trees that are putting out new leaf buds in the spring

Lean soil: Soil, such as clay or sand, that doesn't contain much humus or rich nutrients.

Mulch: This is material laid on the ground around plants to retain soil moisture, insulate the roots during both hot and cold months and reduce erosion and weed growth. *Organic mulches* include bark, wood chips, sawdust, straw, leaves and newspaper; *inorganic mulches* include gravel, stone and ground cloth fabric.

Nectar: The juice of flowers that attracts beneficial insects, hummingbirds and butterflies.

NPK: These three numbers appear on bags of fertilizer (for example, 2-3-2) and refer to the nitrogen, phosphorous and potassium ratio in the mix. N=nitrogen, P=phosphorus, K=potassium (also known as potash).

Depending on the plant and the season, you'll want the percentages to be different.

Organic: Materials originating from a living organism. In a stricter sense, it refers to a method of gardening in which no chemical fertilizers or pesticides are used.

Perennial: An herbaceous plant that lives for more than two years. Perennials usually have one blooming season each year.

Pollination: The spreading of pollen between plants for reproductive purposes and the development of fruit.

Propagation: Plant reproduction, done in three ways:

1) division propagation is taking plant clumps, roots and all, and dividing them into smaller pieces to transplant elsewhere

2) cutting propagation is taking a piece of plant—either a stem or root section—and transplanting it

3) seed propagation means growing plants from seed

Pruning: The trimming of trees and shrubs to take off dead wood, redirect growth or encourage new growth.

Scoring or scratching roots: The loosening of roots when taken from a nursery pot so they won't continue growing in a circle. If tough, they can be cut with a knife. Healthy roots need to stretch out lengthwise.

Scratch-in: This is raking granular fertilizer into the soil, either with fingers, a small rake or hand cultivator.

Soil nutrients: *Macronutrients* are those that are needed by plants in substantial quantities and include carbon, nitrogen, phosphorous, sulfur, calcium, magnesium, and potassium. *Micronutrients* are those that plants need in small quantities and include copper, zinc, iron, manganese, boron, and molybdenum.

Succulents: These are plants with fleshy, thick tissue adapted to storing water. Common succulents include agaves, ice plants and stonecrops.

Suckers: These are growth springing from the roots of trees, some shrubs and roses, often called watersprouts, shoots or canes. Remove suckers, especially from the rootstock of grafted trees and roses to strengthen the main trunk.

Top-dress: This is sprinkling a light layer of compost, mulch or other additives to hold seeds or add nutrients.

Up-pot: Means repotting a plant when it has outgrown its pot and needs to be put into a larger one. Increase the size of the pot and freshen the old potting soil with new soil.

Warm-season grass: Grasses that start greening up when the weather and the soil are warm. They do their growing during warm months and will remain green during the hottest months. They remain dormant in the early spring and will return to dormancy in mid-fall. Also, warm-season grasses do best when planted during warm months.

Water-in: This means using a liquid vitamin, plant food or some other additive with water so it can be readily absorbed by the soil.

Winter protection: This is shelter that keeps the harshness of strong, cold winds and low temperatures off plants that can be damaged by severe climate changes. Some plants such as cacti and succulents need to be brought inside, out of the weather for several months.

Xeric: This is a type of garden or particular plant that tolerates a low moisture growing environment. Terms such as "waterwise" and "water-thrifty" refer to the same concept or type of plant.

Garden Notes

CPSIA information can be obtained at www.ICGtesting.com
Printed in the USA
LVOW06s1010250214

375093LV00001B/216/A